Audio Content Security

Audio Content Security

Attack Analysis on Audio Watermarking

Sogand Ghorbani
Faculty of Computing, University of Technology, Malaysia

Iraj Sadegh Amiri
Photonics Research Centre, University of Malaya,
Kuala Lumpur, Malaysia

ELSEVIER

AMSTERDAM • BOSTON • HEIDELBERG • LONDON
NEW YORK • OXFORD • PARIS • SAN DIEGO
SAN FRANCISCO • SINGAPORE • SYDNEY • TOKYO
Syngress is an imprint of Elsevier

SYNGRESS.

Syngress is an imprint of Elsevier
50 Hampshire Street, 5th Floor, Cambridge, MA 02139, United States

Notices
Knowledge and best practice in this field are constantly changing. As new research and
experience broaden our understanding, changes in research methods, professional practices,
or medical treatment may become necessary.

Practitioners and researchers must always rely on their own experience and knowledge in
evaluating and using any information, methods, compounds, or experiments described herein.
In using such information or methods they should be mindful of their own safety and the safety
of others, including parties for whom they have a professional responsibility.

To the fullest extent of the law, neither the Publisher nor the authors, contributors, or editors,
assume any liability for any injury and/or damage to persons or property as a matter of products
liability, negligence or otherwise, or from any use or operation of any methods, products,
instructions, or ideas contained in the material herein.

British Library Cataloguing-in-Publication Data
A catalogue record for this book is available from the British Library

Library of Congress Cataloging-in-Publication Data
A catalog record for this book is available from the Library of Congress

ISBN: 978-0-12-811383-7

For Information on all Syngress publications
visit our website at https://www.elsevier.com/

Working together
to grow libraries in
developing countries

www.elsevier.com • www.bookaid.org

Publisher: Todd Green
Acquisition Editor: Chris Katsaropoulos
Editorial Project Manager: Anna Valutkevich
Production Project Manager: Mohana Natarajan
Designer: Mark Rogers

Typeset by MPS Limited, Chennai, India

CONTENTS

LIST OF TABLES

LIST OF FIGURES

BIOGRAPHIES

Sogand Ghorbani received her BSc (Software Engineering) from University of Basir Abik, Iran, and her MSc (Information Security) from University Teknologi Malaysia (UTM). She currently conducts research in telecommunications and computer science.

Dr. Iraj Sadegh Amiri received his BSc (Applied Physics) from Public University of Urmia, Iran in 2001 and a gold medalist MSc (Optics) from University Technology Malaysia (UTM), in 2009. He was awarded a PhD degree in photonics in Jan 2014. He has published well over 350 academic publications since the 2012s in optical soliton communications, laser physics, photonics, optics, and nanotechnology engineering. Currently he is a senior lecturer in University of Malaysia (UM), Kuala Lumpur, Malaysia.

CHAPTER *1*

Introduction

1.1 OVERVIEW

These days most of the people are using the Internet to reach so many different purposes, some people are using the Internet just to have fun, and some other people are dependent to it because of their job, e.g., most of the different publishers distribute their own contents such as text or audio or video through the Internet to attract more purchaser and gain more money by having lots of fans from all over the world. But besides all the advantages a big social network community brings, lots of vulnerabilities, threats, and attacks are coming to this area too. Creative content providers are always facing to violation of the copyright of their own work especially through the Internet and online shopping. The phenomenon of Copyright protection, which is an old story, is a big concern of digital content providers nowadays. There are some methods and ways to protect the authentication and integrity of contents through networks like cryptography, watermarking, digital signature, and other ways.

Embedding some information to provide authentication of the owner of the content, e.g., the singer of a song or the related data about some images that can be visible or invisible called watermarking is an old way to hold intellectual copyright protection and some other aims such as "owner identification, proof of Ownership, transaction tracking, and copy control" (Cox and Miller, 2002). But because there are so many threats in the network a watermark is vulnerable to the alternation or even being removed from the host (the content that the watermark is embedded there) by the attackers, the necessity of having new methods to make the watermark more robust is obvious. In addition, regarding to the audio type content, there are numerous attacks that can alter the file and diminish the quality of songs by the way of different types of signal processing. Knowing all types of possible attacks on audio files and analyzing how they can effect on a digital audio content is a vital issue that can help the scientist to propose an appropriate method of audio watermarking by considering all the characteristics of a song and the threats related to it.

Audio Content Security. DOI: http://dx.doi.org/10.1016/B978-0-12-811383-7.00001-5

1.2 PROBLEM BACKGROUND

Nowadays, the kind of ways being used for access to the contents especially digital audios is unlike previous decades. People used to buy the songs and albums from the stores and pay directly, but currently they use different devices to connect to the Internet and find the new albums and download lots of songs usually without paying money rather than old style shopping. The online shopping and publishing requires more security for the data transmission on the Internet, so the need of having security algorithms and methods for distributing data is become more and more important.

During the years of 2000–02 referring to a reporter in *Los Angeles Times*, the music industry has served the media unambiguous statistics in terms of piracy, the act of repetition digital music content to a blank CD, or uploading or downloading them through the Internet. Regarding to numerous newspaper articles, a likely 3.6 billion songs are downloaded every month in the United States unlawfully. This tendency of customers sharing their music rather than purchasing it may be attributable to many factors, including the slow economy. Table 1.1 shows the estimation of audio piracy between the years of 1999–2002.

In 2009 IFPI (International Federal of the Phonographic Industry) had reported that only 5% of downloaded music is legitimate and the rest had been taken and used illegally with no payment to the owner of the content. Among all these unlawfulness and irregular sharing of contents, proof of ownership is a huge concern because everybody who has required knowledge can tamper an audio file and claim that the song belongs to him or her. Audio authentication is significant in content delivery via unreliable channels, e.g., peer-to-peer (P2P) file sharing. Numerous contrarily encoded versions of the original audio might be existent. Differentiating the legitimate diversity of encodings from malicious tampering is the challenge these days (Varodayan et al., 2008).

Table 1.1 Audio Piracy Estimation (Hall, 2002)	
Year	Estimation
1999	The music industry estimated that one in four compact disks of new music was actually an unauthorized copy. And since this year, ownership of CD burners has nearly tripled
By the end of 2001	It was estimated that as many CDs were burned and copied as were bought. In Europe, blank CDs are outselling recorded CDs (although these blank CDs might have also been purchased for legitimate reasons, such as to backup personal computer files)

"Digital audio watermarking has been recognized as a helpful way with dealing with the copyright protection problem in the past decades. Although digital watermarking still faces some challenging difficulties for practical usage, there are no other techniques that are ready to substitute it" (Lu, 2005). "We believe that informed watermarking offers significant near-term improvements. While proposed codes for informed embedding are computational efficient, they are not robust to volumetric scaling" (Cox and Miller, 2002). Later approaches can be extended with cryptographic approaches like digital signatures. To allow different security levels, we have to identify relevant audio features that can be used to determine content Manipulations (Steinebach and Dittmann, 2003). There is a list of ways that can make an alteration to an audio file in Fig. 1.1.

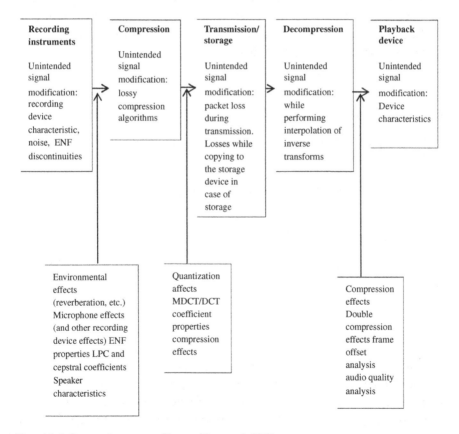

Figure 1.1 Audio processing system architecture (Gupta et al., 2012).

1.3 PROBLEM STATEMENT

Although watermarking technique is a good way to preserve copyright information and it is using for all contents such as text, image, and video, for audio type contents it is not so public yet. There are a lot of methods proposed by scientists to do audio watermarking but the problem is none of them are focusing on a particular genre of music, while various genres of music have different characteristics that can have effect on the method and each genre need be tested particularly.

1.4 PURPOSE OF STUDY

The core driver of this study is to analyze the effects of different types of audio signal processing (attacks) on audio files and to characterize them into particular groups of songs. The major focus is on the effective parameters on the evaluation of an audio watermarking method that compare the original audio with the watermarked one to reveal proportion of their similarity.

1.5 OBJECTIVES OF STUDY

The subsequent objectives are indicated to accomplish the intent of the study:

1. To study the current watermarking techniques for audio
2. To analyze the possible signal processing on audio files and observe their effects on the embedded watermark and the audio itself regarding to different genre of music
3. To correlate two parameters of audio structure and watermark (size ratio and sampling rate) on PSNR

1.6 SIGNIFICANCE OF STUDY

This study deliberates a series of useful results from various tests on security and performance measures on a specific audio watermarking method that can be very useful for further studying in this area. The way of analysis is so clear and organized regarding to security and performance dealings.

1.7 SCOPE OF STUDY

The major focus of this study is copyright issues on audio files and the ways that they can be protected. The boundaries that are covered in this study are representing as follows:

1. The study will discuss different watermarking methods for audio
2. The requirements of a high-quality audio file which are security, capacity, imperceptibility, robustness, and time consumption are being analyzed
3. Forty audio file samples would be chosen from four different categories (Rock, Classical, Ambient and Pop) to place the watermark inside
4. Different types of signal processing (compression, resampling, filtering, and delay) would be run on all the audio files
5. Watermark extraction would be done to see whether the watermark can be extracted after attacks successfully or not
6. Discussion about different ways of compares the watermarked audio with the original host to measure methods performance
7. Measuring the selected method performance using PSNR
8. Correlation between different parameters with PSNR

1.8 ORGANIZATION OF THE REPORT

This study is separated into 6 chapters that so far is introduced and defined concisely the structure of the study would be organized as following statements:

Chapter 1, Introduction, describes briefly about the overview of the project and understanding of the project problem background then the statement of the problem. It furthermore comprises the project scope, purpose of the study, and the objectives.

Chapter 2, Literature Review, discusses about the general definition of creative content and copyright issues then focuses on the audio file and its characteristics and requirements, and introduces the watermarking methods. Besides it has an explanation of different audio signal processing and attacks and the ways to see their exact effects on audio files and watermarks. Chapter 3, Methodology, consists of the methodology of this research in details to clarify what exactly would be done in this study and introduce the process of leading this project step by step in three different phases of the framework.

Chapter 4, Attack Analysis, contains explanations of the design and implementation of this study. In this chapter there would be heaps of tables and figures to show the result of various testing.

Chapter 5, Evaluation Analysis, explains different methods of comparison between two different audio signals would be introduced and one of them is selected for the further processing of this project. Finally, Chapter 6, Conclusion, analyses and abstracts the whole project conclusions and suggests some recommendations.

CHAPTER 2

Literature Review

2.1 INTRODUCTION

Nowadays people are using lots of contents to have a better life and gain more information from different types of them. Audiovisual media like film, television programs, documentaries, news and blogs, etc., cover lots of areas of people's interests and it is a very good way to share their knowledge, art, and creativity because everyone can access and use it easily. Also there are various types of online services like Music, radio, Games, or even online publishing like books, newspaper that they can be in the form of educational content, cultural information, and so on. Although some of these resources are free to access and people are allowed to use them without paying money, most of them has some limitations and needs to be paid before download or used. Unfortunately, there are many ways to bypass the law and misuse of this availability especially through the Internet. One of these ways has been paying just once for content and shares it illegally among your friends or even sells it. Above all these economic lost, there is an opportunity for the sabotages to manipulate the contents and claim that this content belongs to them. Because of these reasons intellectual property copyright is a very important issue and watermarking is one of the useful copyright protection ways.

This chapter is about the importance of creative content copyright and the main focus is on the audio type content issues. The history of watermarking methods in audio will be introduced here and the current methods of audio watermarking would be discussed too. Farther, details of different types of audio files and their special characteristics are explained here; also this chapter clarifies how an audio can be changed from an analog signal to a digital one. After those different methods of watermarking and different attacks on them would be discussed by the end of this chapter, there is a part about how to evaluate and compare host and watermarked audio.

Audio Content Security. DOI: http://dx.doi.org/10.1016/B978-0-12-811383-7.00002-7

2.2 CREATIVE CONTENT

Creative content refers to every possible types of content which belongs to one or a group of creators where creators consist of every possible cortex of society who has new ideas about every imaginable thing in the world that could be shared with other people. Because of this fact that people have different knowledge and viewpoints which qualifying those to create their own content by using of every available Media like text, video, image, or audio and after producing their own unique content, they want to share it among the people, exactly at this moment the importance of copyright issues is concerned because no one likes to lose their own authority and sense of ownership of the product that originally belongs to them.

2.2.1 Creative Content Copyright

As a matter of fact, this is very obvious that free copying and distribution of creative contents is so easy these days as a consequence of convergence of three special parameters, which are widely available network connection, personal computers and mobile phones, and diversity of digital contents. The result has been extraordinary piracy of content, expressive music, which the music industry complains for momentous financial losses each year (Kevin, 2007).

2.3 DIGITAL AUDIO

A format for storing digital audio data on a computer system is called an audio file format. This information can be stored in two ways: uncompressed, or compressed to diminish the size of the file. It is possible that the audio file is a raw bit stream; however, it is typically a container format or an audio data format with defined storage layers. There is significant differentiation among a file format and an audio codec. All compression formats consume a codec of several types, which is a process or a software program that compresses and then decompresses an audio signal constructed on a quantity of rules and assumptions about how we perceive music. And also it is essential to distinguish between codecs and wrapper (or container) formats. Codec denotes to the algorithm used to encode and decode the audio data in binary form; "wrapper" defines the container format for this raw data, which may contain headers to describe encoding settings, as well as other content—such as artwork or even video data (with certain wrappers)—and expressive

Table 2.1 Audio File Categories			
Audio File Format	**Uncompressed**	**Lossless Compression**	**Lossy Compression**
Characteristics	• No compressions on the Best choice for quality • Big size • Silence is included • Metadata included	• Eliminating unnecessary data • Recreated uncompressing data • Smaller file • The silence takes up no space at all when being encoded	• Greater reduction than lossless • Simplifying the complexity of data • Reduction in quality • Degree of compression measured in bit rate
Suitability	• Archiving and supplying audio at high resolution due to their correctness • Suitable when working with audio at a professional level	• Used in many applications For example, in the ZIP file format • Frequently used as a component within lossy data compression technologies • Most often used for archiving or production purposes	• The most generally used to compress multimedia data (audio, video, and still images), particularly in applications such as streaming media and Internet telephony

metadata (Atoum et al., 2011). Different types of wrapper will be denoted by different file extensions, such as .wav, .aiff, .mp3, etc. The data itself is kept in a file with a precise audio file format while a codec makes the encoding and decoding of the raw audio info. Though most audio file formats support only one kind of audio data (produced by an audio coder), a multimedia container format (as Matroskaor AVI) could support multiple kinds of audio and video data. There are three main groups of audio file formats as shown below (Yershov and Karpelcev, 2011, Yu et al., 2011). A categorization of the different types of audio file is illustrated in Table 2.1, and in each category there are lots of file extensions with their own specific characteristics, in Tables 2.2–2.4 the most popular types belonging to each category is shown.

2.3.1 Convert a Sound to an Audio File

Audible variation in the air pressure is called sound and the thing that converts them to a variable voltage is microphones. A process called Analog to Digital conversion is used to convert these voltages into a sequence of digits to denote a sound digitally. The produced audio data containing these numbers is in the form of PCM (pulse code modulation) format basically but in order to compress it lots of other formats are available. The continuous voltage as a consequence of the air pressure variation created by a microphone is extended in two dimensions as shown in Fig. 2.1.

Table 2.2 Uncompressed Audio File

Uncompressing Format

File Extension	Advantages	Disadvantages	Company Creation
WAV Waveform Audio File Format	• A flexible format • High-quality audio • Scarce limits that extensions to the format have been advanced to cover them	• Its incapability to hold any metadata • File size is also limited to 4 Gb	Microsoft
BWF Broadcast Wave Format	• Functionally matching and cross-compatible with WAV • Contains an additional header file which covers (metadata) about the audio and synchronization info • Typically a BEXT chunk or (more newly) iXML chunk	• 4 GB size limit for the same reasons as WAV	European Broadcasting Union
AIFF Audio Interchange File Format	• Works similarly to WAV but uses a different method of dividing the PCM data into manageable chunks • Widely available free codecs for all platforms • AIFF is the native format for audio on Mac OSX	• The same as all uncompressed audio files	Apple and Amiga

Table 2.3 Lossless File Format

Uncompressing Format

File Extension	Advantages	Disadvantages	Company Creation
ALAC Apple Lossless Audio Codec	• Very fast decoding • Open source • Hardware support • Streaming support • Tagging support (QT tags)	• Limited software support • No hybrid/lossy mode	Apple
FLAC Free Lossless Audio Codec	• Open source • Very fast decoding • Fast encoding • Hardware support • Software support • Error robustness • Streaming support	• No hybrid/lossy by mode	Developed by Josh Coalson
APE Monkey's Audio	• Open source • High efficiency • Good software support • Simple and user friendly. Official GUI provided • Java version (multiplatform) • Tagging support	• No multichannel support • No error robustness • No hybrid/lossy mode • Limited hardware support • Higher compression levels are extremely CUP intensive	Matthew T. Ashland

Table 2.4 Lossy File Format

Uncompressing Format

File Extension	Advantages	Disadvantages	Company Creation
MP3 MPEG Audio Layer-3	• Very popular • Compatible with most portable music players • Eliminate parts of the music that the human ear finds tough to hear. • ISO standard • Fast decoding • Anyone can create its own execution	• Problem cases that trip at all transform codecs • Slow encoding (using Lame VBR) • Sometimes, maximum bitrate (320 kbps) isn't enough • No multichannel implementation	• The German company Fraunhofer-Gesellshaft
OGG Vorbis	• Open and free • Claimed best quality lossy audio codec (at certain bitrates) • At a beta stage, Ogg Vorbis has been already among the leaders of lossy audio compression and improvements are occurring steadily.	• Limited official development (third-party development is always encouraged) • Present implementations are more computationally demanding to decode than MP3	• Xiph.Org Foundation
AAC Advanced Audio Coding	• Designed as an improved-performance relative to MP3. Used by iTunes • Transparent quality at "archive", very near transparent at "extreme" • Usable in low-delay streaming • Many sampling rates (8000–96,000 hy), up to 256 kbits/s per channel, up to 48 channels	• Inherent problems as a transform codec • Slow encoding • Increased complexity	• Developed by a group of companies containing AT&T Bell

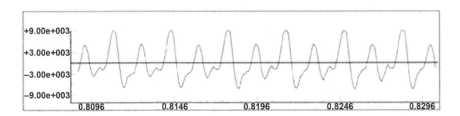

Figure 2.1 Continuous waveform (http://www.billposer.org/).

To represent these continuous signals by a digital system like a computer, the following procedures should be done, measure the signal at a finite set of discrete times, this is called sampling. Make use of a finite number of discrete amplitude levels. This is called quantization.

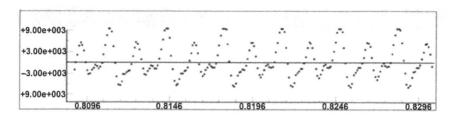

Figure 2.2 Sampled and quantized representation (http://www.billposer.org/).

Resolution stands for the number of levels used which is commonly stated in bits as the base-2 logarithm of the real number. For example, a system with a resolution of 4 bits makes use of $2^4 = 16$ levels. In general, the quality of a digital audio content depends on the sampling rate and the resolution of that. Are solution of 16 bits and a sampling rate of 44,100 samples per second is a CD-quality. Fig. 2.2 shows the converted representation of a continuous waveform.

2.3.1.1 Effective Parameters on Audio Quality
There are some important factors that have effect on audio quality:

Sampling rate: A process which splits the audio wav files into different segments and storing their shape and any other attributes like noise in a digital format, i.e., zero and one, is called sampling (Jackson, 2013). The number of samples in each unit of time (often seconds), which is reserved by continuous signal to make a discrete signal is called the sampling rate or sample rate and is defined by the Nyquist Sampling Theorem. The greater the sampling rate, the more perfectly the tested signal will denote the original signal.

Bit Rate: It defines in every second how many bit is conveyed or processed; the higher the bit rate, the higher the quality. The bit rate can change dynamically and with all the other parameters equal, quality depends on higher bit rate (You et al., 2013).

Channels: A solo stream of sound, such as that from a normal monaural recording, establishes one channel. Stereo requires two channels. Quadraphonic music requires four channels (Deschamp et al., 2012).

2.3.2 Audio Copyright
Some important factors of an audio production should be protected from any type of manipulation. Whether this audio is a sound, music,

or voice chat, it is required to capture some basic information. In recent years a fast development in network communications had been accrued and as a result of that the delivery of Audio content especially music has become more effective and informal than ever before. But a parallel grows on the unlawful distribution of copyright-threatened contents has happen too which is a serious concern (Li et al., 2010).

2.4 WATERMARKING

Watermarking is an embedding system which has the ability to embed a signal called embedded signal or watermark into another signal named host signal in a manner that no grave ruin to the host has been happening by the embedded signal. At the similar while it is essential that the embedding be robust to common degradations to the merged host and watermark signal, which in some applications result from intentional attacks "Ideally, whenever the host signal survives these degradations, the watermark also survives."(Chen and Wornell, 2001). Regarding to the main definition of watermarking, it is needed to distinguish it with the definition of steganography although in both terms there is some information that is being embedded into a host, the purpose of them is totally different. Steganography is a way to cover a secret message by hiding it to other content with no specific relation between them likewise the watermarking used to put related information of a content owner or publisher to protect the copyright of it.

2.4.1 Audio Watermarking

The aim of digital audio watermarking is to provide the artistic and creative work owner the capability to verify that the work has belonged to them. Besides copyright protection purposes, an audio watermark has other useful effectivenesses such as consisting a metadata which is additional description about the artist or the content or even you can add a URL address in the audio file to tell the purchaser where to get more information about the song or download the full album. "Digital Watermarking" is the technique of embedding some information into multimedia content by modifying the media content slightly. The information, called a watermark, can be extracted from the market media content (Lian, 2009). General architecture of a watermarking system is shown in Fig. 2.3.

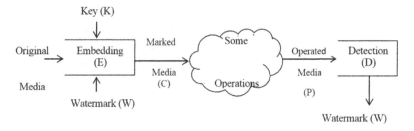

Figure 2.3 General architecture of watermarking system.

2.4.1.1 Fundamental Properties of Audio Watermarking

A definition of watermarking has been given so far, now it is time to have a look at core properties that comprise a watermark. Although an ideal watermark should have all the listed requirements, there are some challenges practically that restricts watermark companies. According to the International Federation of Phonographic Industry, it is necessary that audio watermarking meet several requirements such as below:

Imperceptibility: The most important requirement of audio watermarking is that the quality of the original signal has to be retained after the embedding of watermark. The digital watermark should not affect the quality of the original audio signal after it is watermarked. Referring to the terms of audio, the word Inaudibility is being more suitable.

Robustness: The embedded watermark data should not be removed or eliminated by using common audio signal processing operations and attacks. The detection rate of watermark should be perfect. Noise reduction and Normalization are two samples of common operation on the audio. A watermark is robust when it remains unmovable and also readable after all these operations or even attacks occur.

Effectiveness: It refers to the likelihood of recognition of a watermark by detector immediately. The time that a watermark is not successfully recovered by a decoder is called that a false-negative error has happened.

Capacity: It refers to the number of bits that can be embedded into the audio signal within a unit of time. A user should be able to alter the amount of information embedded depending upon the applications.

Security: It implies that the watermark can only be detected by the authorized person. Speed: The watermark embedding and extracting processes have to be fast enough depending upon the application (Yu et al., 2011; Patil and Chitode, 2012).

2.4.1.2 Types of Watermark
There are so many types of watermark with specific characteristics but regarding to Lu (2005) Table 2.5 is provided here to show five types of them.

2.4.1.3 Watermarking Techniques
By looking at Lu (2005), Table 2.6 is mentioned to specify different types of techniques being used for watermarking.

Table 2.5 Types of Watermark (Lu, 2005)	
Type	Decryption
Robust watermark	A watermark which is robust against attacks because even if the existence of it is known it is so hard for the attacker to manipulate it as there is secret key inside the watermark
Fragile watermark	The watermarks that the robustness of them is so limited. The usage of them is to detect the alteration of the data instead of conveying irremovable information
Perceptible watermark	They are easy to find by the consumers, usually is seen in the images or logos, and also in audio they exist to overlay on top of the musical work to discourage stealing
Bit stream watermark	These types directly embed into compressed audio or video and being used in the environments that have limitation of space like the Internet
Fingerprinting and labeling	The fingerprint is referring to the time that the watermark consists of the information that has a unique code to identify the recipient for confidential purposes. In labeling the data embedded is a unique data identifier like a library retrieving

Table 2.6 Techniques of Watermarking	
Technique	Explanation
Amplitude modification	Also known as least significant bit substitution and the information are being encoded into them using two ways
Dither Watermarking	It is a noise signal that is added to the input audio signal to provide better sampling of that input when digitizing the signal
Phase Coding	It embeds information on the original discrete audio signal by introducing a repeated version of component of the audio signal
Echo Watermarking	It works by substituting the phase of the original audio signal X with one of two reference phases each one encoding a bit of information
Spread Spectrum	The core idea is to embed a narrow-band signal into a wide-band channel. It offers a possibility of protecting the watermark secrecy by using a secret key to control the pseudorandom sequence generator that is needed in the process

2.4.2 Audio Watermarking issues

So far it has been discussed that audio copyright can be protected by watermarking; also it is said that watermarking for this kind of content is not as much popular as for the other types. One of the reasons is its vulnerability in face of different types of intentional or unintentional attacks. Watermarking requirements are various and each one could be compromised by specific signal processing and attacks, e.g., in the first step the embedded watermark can cause a significant effect on audio which leads to be heard easily by human.

2.4.2.1 Conversion Issue

Most of the times the original format of audio files which is wav and has a lot of qualitative factors and huge size is being converted to the MPEG Layer III (MP3) and it will cause some part of the file lost and brings challenges for audio watermarking produces to design a system which can resist on this type of attack. Fallahpour and Megias (2012) intend a high-capacity audio watermarking algorithm in the logarithm domain based on the absolute threshold of hearing of the human auditory system which makes this scheme a novel technique. The experimental results show that the method has a high capacity (800–7000 bits per second), without significant perceptual distortion Objective Difference Grade (ODG is greater than -1) and provides robustness against common audio signal processing such as added noise, filtering, and MPEG compression (MP3). Fallahpour and Megías Jiménez (2012) proposed a novel high-capacity audio watermarking algorithm to embed data and extract them in a bit-exact manner by changing some of the magnitudes of the FFT spectrum. The experimental results show that the method has a high capacity (0.5–2.3 kbps), without significant perceptual distortion (ODG is about 1) and provides robustness against common audio signal processing such as echo, added noise, filtering, and MPEG compression (MP3).

2.4.2.2 Resampling Issue

Resampling is one of the common signal processing in which the audio signal could be down sampled, up sampled, and over sampled and each one can cause some changes in the audio frequencies. Lei et al. (2012) proposed a new and robust audio watermarking scheme based on lifting wavelet transform (LWT) and singular value decomposition (SVD). Specifically, the watermark data is efficiently inserted in the coefficients of the LWT low-frequency subband taking advantage of both SVD and quantization index modulation. Experimental and analysis results demonstrate that the proposed LWT-SVD method is not only robust against both general signal processing attacks and desynchronization attacks but also achieve a very good tradeoff between robustness, imperceptibility, and payload. Kaur and Kaur (2013) implemented a variety of audio watermarking methods to test and evaluate them using blind detection techniques and in conclusion reveals that previously techniques focused on only improving robustness against common signal manipulations and improving perceptual quality, but recent techniques also focus on desynchronization attacks or others along with common attacks.

2.4.2.3 Filtering Issue
An audio filter is a frequency-dependent amplifier circuit, working in the audio frequency range every audio file can pass from a high-pass filter or low-pass filter. Chen et al. (2013) present an adaptive audio watermarking method using ideas of wavelet-based entropy (WBE). The method converts low-frequency coefficients of discrete wavelet transform (DWT) into the WBE domain, followed by the calculations of mean values of each audio as well as derivation of some essential properties of WBE. A characteristic curve relating the WBE and DWT coefficients is also presented. The foundation of the embedding process lies on the approximately invariant property demonstrated from the mean of each audio and the characteristic curve. Besides, the quality of the watermarked audio is optimized. In the detecting process, the watermark can be extracted using only values of the WBE. Finally, the performance of the proposed watermarking method is analyzed in terms of signal-to-noise ratio, mean opinion score, and robustness. Experimental results confirm that the embedded data are robust to resist the common attacks like resampling, MP3 compression, low-pass filtering, and amplitude scaling.

2.4.2.4 Delay Issue
The time interval between a signal and its repetition is called Delay which is most often applied as a send effect, rather than as an insert. This approach not only conserves processing power when applying the same delay to multiple sources, but it allows treating the delayed part of the signal, separately from the original, with extra processing such as EQ or distortion and this brings more creative freedom. One of the

most common delay effects on audio in echo. Wang et al. (2012) measured the accuracy of reconstructed watermark from audio signal after watermark attacks by various signal processing methods. The experimental results indicate that the proposed zero-watermarking method has better robustness as compared with methods proposed in literature. The proposed scheme is efficient and can be applied in the protection of intellectual property rights of audio signals.

2.4.2.5 Desynchronization Issue
Nowadays resisting on this type of attack is being more important than before for audio watermarking developers. Desynchronization refers to a process causing an absence of synchronization, the relation that exists when things occur at unrelated times; as the stimulus produced a desynchronizing of the brain waves, Peng et al. (2013) proposed a method for audio watermarking which is robust against desynchronization attack and is based on kernel clustering.

2.5 ATTACKS ON AUDIO

There is a lot of signal processing that can lead a negative effect on audio songs to specializing these various attacks in particular groups (Table 2.7).

2.6 EVALUATION METHOD

There are various methods and algorithms which are based on mathematics to evaluate a signal and compare tow audio files. To choose the best one, it is mandatory to know what is needed exactly to compare and what is going be the results Table 2.8 has a brief explanation on different methods.

Table 2.7 Attacks on Audio Watermarking	
Removal Manipulations	**Misalignment Manipulations**
Addition of multiplicative and additive noise	
Filtering (Low–High...)	Fluctuating time and pitch scaling
Lossy compression like mp3	
Noise reduction	
(D/A) and (A/D) conversion	Cropping and insertion of samples
Changing the sampling rate	
Collusion and statistical attack	

Table 2.8 Evaluation Methods		
Method	Terminology	Definition
PSNR	Peak signal-to-noise ratio	It is an engineering term for the ratio between the maximum possible power of a signal and the power of corrupting noise
SSIM	Structural similarity	It is a method for measuring the similarity between two images or two songs
SNR	Signal-to-noise ratio	The signal-to-noise ratio (SNR) is the measure of the maximum output voltage compared to the integrated noise floor over the audio bandwidth, expressed in db

2.7 PROJECT CRITERIA

After gathering all the information, this study is planned to offer an effective consideration for audio watermarking algorithms based on precise and actual analysis on attacks which are launched to different genre of music songs to see their robustness and compare them to each other also one of the most sensible part of audio watermarking which is evaluation of its proficiency is discussed.

CHAPTER 3

Methodology

3.1 INTRODUCTION

In Chapter 2, Literature review, the principles of study and conventional and new methods in audio watermarking have been recognized and discussed. Also an introduction about different types of audio watermarking has been done with referring to previous works who evaluate their work with testing it via launching some attacks. The discussion of the methodology used in this study is organized in this chapter. The research framework of this study is based on the main three objectives of this study and each phase aimed to meet each goal particularly. The three different phases are as follows: literature review and experiment infrastructure, attack analysis and results, evaluation analysis and result. In each phase there are different steps that should be followed respectively. The details related to each phase are discussed in this chapter too. The implementation of the framework and information of that are discussed in further chapters.

3.2 RESEARCH FRAMEWORK

The study is separated into three phases and every phase brings the projected output which suits the necessity input for the succeeding stage. The indication of the research framework for the planned technique is presented in Fig. 3.1.

3.2.1 Research Framework Phase One

In the first phase of this study the principles of audio structures and watermarking methods are discussed and presented as a literature review. The literature review is divided into different sections and each section covers a basic part of the study. Explanation of creative content and audio files and their vital issues on protecting the copyright is provided first and the ways to protect the copyright are presented next. The audio watermarking method is introduced as a good way to

Audio Content Security. DOI: http://dx.doi.org/10.1016/B978-0-12-811383-7.00003-9

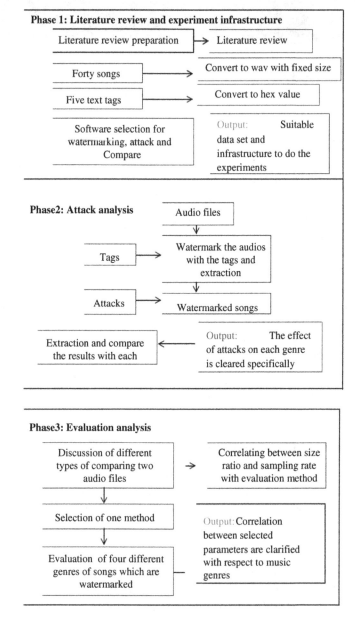

Figure 3.1 Research framework.

protect copyright issues. Further sections in literature review are focusing on different aspects of audio watermarking such as the history and the main methods of it, major characteristics that a method should have and different types of attacks that an audio watermarking method can face, and also the ways to evaluate an audio watermarking method. After completion of literature review, an integrated dataset is select. The dataset consists of 40 songs from 4 different group of music and 10 songs for each type. Each genre has different characteristics and the songs have different range of frequencies and it brings challenges to the embedding and extraction process. The music genres that are selected for this project are rock, pop, classical, and ambient. All the songs are converted to wav file using Aiseesoft MP3 WAV Converter. Because the size of all songs should be the same, a process for making equal size for files is done using Audacity software. After that five different text tags with five different sizes are produced and converted to Hex Values using Cryptools. The reason to have different sizes of tags is to test the performance of the watermarking method in face of small or big tags. In the last stage of phase one after preparing all audio data and text type tags, the audio watermarking method should be selected. The audio watermarking is not as famous as image watermarking and there are a few applications that are capable to do the watermarking for audio files. EYM Audio Watermarker is selected for implementing this study. This application is working only with Wav files and it is not possible to watermark any other type of songs like mp3 format or the others with it. The audacity software is a very good application for working with audio files and editing them; in this study all the attacks are launching to the watermarked songs using this app. Besides the possible method for evaluation of watermarking are assessed. Finally the needs to run the projects are satisfied and the next phase experiment will start.

3.2.2 Research Framework Phase Two

In the second phase of this study the selected watermarking algorithm is used to practice watermarking on 40 different songs which belongs to 4 different genre of music. The process is repeated five times as there are five different tags. After preparing all the watermarked songs, the process for extraction is done to see how many percent of songs in each music genres can retrieve the watermark successfully before launching any attacks. The result of this stage is very useful to see the difference between rock, pop, classical, and ambient songs regarding to success of extraction in a same setting of process. Also the results will show the effects of various sizes of tags on the watermarking method. The proportion of success for each category of songs are kept and shown in charts and diagrams and tables to be compared with further results. Next step in this study is to run the attacks on watermarked audio songs, there are lots of attacks and signal processing on audio files that could lead to diminish the audio quality and also removal or tamper of the watermark inside the audio. The selected attacks to run in this study are mp3 conversion, resampling (upsampling and down sampling), filtering (low-pass filter and high-pass filter), and echo. These attacks are launched to each genre separately and after that the attacked watermarked song will go through extraction. All the extraction results after attacks, i.e., whether success or failure, are documented carefully to be used for analysis. The analysis in this part is divided into two main parts. One aspect of analysis is to analyze the effect of different attacks on a particular genre of music. And the other aspect is to compare the robustness of different songs from different genres of music on a particular attack. For the first goal of analysis, the study will reveal that every genre of music could be vulnerable to some attacks more than the others. In this part the examiner tries to show that success and failure of an audio watermarking algorithm is not just dependent on the method for embedding and extraction, it is also reliant on the type of music that is using as a host. And the aim for the second aspect of analysis is to compare different genres of music in face of a particular attack. Finally, all the tables and diagrams are summarized to one section and the last result of the analysis is depicted in this phase and after that next phase will start.

3.2.3 Research Framework Phase Three
In the last phase of this study a discussion related to the evaluation methods of audio watermarking is conferred. The basic concept of evaluation is to see how much the watermarked song is similar to the original song. The more similar they are, the better algorithm it is. Peak signal-to-noise ratio (PSNR) is a known way to evaluate the watermarked audio and the original one. PSNR is an engineering term for the ratio between the maximum possible power of a signal and the power of corrupting noise that affects the fidelity of its representation. Because many signals have a very wide dynamic range, PSNR is usually expressed in terms of the logarithmic decibel scale. There is a good application named Diffmaker that can subtract the watermarked audio from the original one and show the remains as a new audio file. If the difference is high, the new audio is a strong noise and can be hear easily but if there is a little difference, then the noise can be heard

hardly. In this phase the study estimates PSNR of the method for different genre of music using this application. There are two purposes of doing this process, first of all the effect of size ratio and sampling rate on the PSNR is measured and second the result reveals how much difference is between different genres of music regarding to the PSNR.

3.3 RESEARCH TOOLS USED

The software and hardware requirement of the personal computer that is used is determined in this section. First the system hardware specifications are identified:

- Computer type: Laptop ASUS
- Processor: Intel (R) Core i7 q740 @1.73GH
- Installed Memory: 4.00 GB
- System Type: 64-bit operating system
- Cash: 6 MB
- Samsung Smart Phone S III Mini

Then the system software is illustrated as follows:

- Operating System: Windows 7-Service Pack 1
- Aiseesoft MP3 WAV Converter
- EYM audio watermarker
- Audio Diffmaker
- Audacity
- Microsoft office
- EndNote

3.4 SUMMARY

In this chapter the research methodology has been deliberated and defined the software and hardware tools that are used in this research, also research framework with three phases is illustrated to achieve research objectives that are mentioned in Chapter 1, Introduction, and the operational framework in three parts has been shown to guide this study.

Attack Analysis

4.1 INTRODUCTION

In this section the discussion about findings of the impact of attacks on an audio watermarking method which is using Fast Fourier Transform (FFT) is discussed in details. The process is based on effects of different attacks on four different genres of music (Rock, Pop, Classical, and Ambient). The results are shown in a clear way for every specific type and by the end of this chapter there is an integrated comparison between all kinds of songs and their resistance on the attacks. Based on the research framework from Section 3.2 of Chapter 3, Methodology, discussion of the first phase is presented in this chapter.

4.2 ATTACK ANALYSIS WATERMARKING ALGORITHMS

In this chapter first of all 40 songs are chosen to be watermarked using EYM software which is using dual watermarking method and FFT for embedding and reveres for extraction. The project focuses on the different types of attacks and songs to find their relationship before using a method for audio watermarking. However, from the achieved results of this study the developer benefits in order to make his application less vulnerable to other competitor or attacker. This chapter is aimed to formulate problems in audio watermarking generally. Tools, dataset, and specifications involved in the analysis are discussed in detail in Section 4.2.1.

4.2.1 Results on the Analysis of Audio Watermarking

In this analysis, a variety of songs that are basically consisting of four different genres of music were used. First group is Rock song which is a combination of the rhythm, instruments, vocals, and attitude. Second one is Pop songs that the basic form for them is the song and usually a song consisting of verse and repeated chorus. The third one is Classical music and is performed on certain instruments, such as the piano, trumpet, or human voice, and the fourth one is Ambient music

Audio Content Security. DOI: http://dx.doi.org/10.1016/B978-0-12-811383-7.00004-0

which are subtle instrumentals mostly used as background music but is also meant to be listened to. It focuses more on sound than melodic form. Throughout the analysis, several problems were encountered during embedding and recognition process and it is documented as part of the results and comparisons. After conducting watermarking process, five different types of attacks (MP3, resampling (down sampling and upsampling), low-pass filtering and high-pass filtering, and echo) is done to the watermarked audios and through extraction part the impact of them on the files is revealed whether the watermark can be extracted successfully or it cannot. The detail about each one would be explained in Table 4.1 and the key point is that although the selected songs are from different categories of music genre, they are the same in basic factors (size, bitrate, and length) as they need to be comparing to each other.

4.2.2 Tag Generation

In this project five samples of possible tags are embedded into the songs to see how many bytes can be effectively extracted from the watermarked audio and after that the attacks would be launched to the songs to see how much differentiation is between small tags and the large ones regarding to their successful extraction. Data payload is arbitrary in this method, so five different size of tags from 5-bytes to 25-bytes is generated and a process of conversion from text to hexadecimal values is done then the tags are turned into a data packet that includes some synchronization, error checking, and the optional 12-bit timestamp information. For instance, carrying a payload of simply 1 byte (8 bits) with a noncompulsory timestamp will generate a data packet that becomes repeated at a 1.7 seconds interval throughout the audio stream (i.e., the equivalent of a 4.7-bps data rate).

On the other hand, assigning a payload of 8 bytes (64 bits) without timestamp will create a data packet that gets repeated at 3.4-seconds interval throughout the audio stream (i.e., the equivalent of an 18.8-bps data rate). But thinking of the data payload effectiveness in terms of its encoding data rate can be confusing. Indeed, the audio stream can be thought of as a tempestuous and intermittent data channel. Not all embedded watermarks will go through, even if the audio material is not degraded. This is why the same data gets repeated throughout the audio stream, allowing the decoder to use an expected redundancy in order to retrieve that payload despite the intermittence of the data channel.

Table 4.1 Dataset				
Name	**Genre**	**Size (MB)**	**Bit Rate (kbps)**	**Length**
The rolling stones (01)	Rock	33.1	1536	3:00
Clocks (02)	Rock	32.9	1536	2:59
Blueberry Hill (03)	Rock	33.5	1536	3:02
Hotel California(04)	Rock	32.9	1536	2:59
Runaway (05)	Rock	32.8	1536	2:59
The platters (06)	Rock	32.9	1536	3:00
Preview (07)	Rock	32.4	1536	2:27
Stayinglive (08)	Rock	33.0	1536	3:00
Queen (09)	Rock	32.7	1536	2:58
I got you Babe (10)	Rock	32.9	1536	2:59
Your Love Is My Drug (01)	Pop	32.9	1536	2:59
Please don't go (02)	Pop	32.8	1536	2:59
Today Was A Fairytale (03)	Pop	33.2	1536	3:01
Jar of hearts (04)	Pop	32.9	1536	3:00
Mine (05)	Pop	32.8	1536	2:59
All I Ever Wanted (06)	Pop	32.9	1536	2:59
This Afternoon (07)	Pop	33.0	1536	3:00
Hold my hand (08)	Pop	32.9	1536	2:59
Dynamite (09)	Pop	32.9	1536	3:00
If we ever meet again (10)	Pop	33.0	1536	3:00
Morning (01)	Classical	32.9	1536	2:59
Pachelbel (02)	Classical	32.8	1536	2:59
Piano concerto (03)	Classical	33.1	1536	3:01
Flower Duet (04)	Classical	33.2	1536	3:01
Debussy (05)	Classical	33.0	1536	3:00
Violin Concerto In E Minor (06)	Classical	32.9	1536	2:59
Gorecki (07)	Classical	32.9	1536	2:59
lute And Harp Concerto (08)	Classical	33.0	1536	3:00
Song to the Moon (09)	Classical	33.0	1536	3:00
Misha Mishenko (10)	Classical	32.9	1536	2:59
Arcanozero (01)	Ambient	33.0	1536	3:00
Sleep_Walking Experience (02)	Ambient	33.0	1536	3:00
In This On Me (03)	Ambient	32.8	1536	2:58
The_Lake (04)	Ambient	33.0	1536	3:00
Terrestrial Jerusalem (05)	Ambient	33.4	1536	3:02
Muhmood (06)	Ambient	33.0	1536	3:00
Bleeding_Kansas (07)	Ambient	33.1	1536	3:01
Satellite (08)	Ambient	33.0	1536	3:00
I Just Want To Be Your Everything (09)	Ambient	32.9	1536	3:00
The Black Heart Rebellion (10)	Ambient	33.0	1536	3.00

Table 4.2 List of Tags		
Tag	Hex Value of the Tag	Size (Byte)
A sogig	736f676967	5
B sogandutm2	736f67616e6475746d32	10
C sogi1988utm2013	736f67693139383875746d32303133	15
D sogandghorbanigradut		20
	736f67616e6467686f7262616e69677261647574	
E sogandgraduate20131230utm	736f67616e64677261647561746531746532303133133132333075746d	25

Table 4.3 Information of Embedding Process for Rock Songs					
	Tag A/Number of Packets	Tag B/Number of Packets	Tag C/Number of Packets	Tag D/Number of Packets	Tag E/Number of Packets
Song1	146	90	64	50	42
Song2	146	88	64	50	42
Song3	148	90	66	52	42
Song4	146	88	64	50	42
Song5	146	88	64	50	42
Song6	146	88	64	50	42
Song7	144	88	64	50	42
Song8	146	90	64	50	42
Song9	144	88	64	50	40
Song10	146	88	64	50	42

From this viewpoint it is easier to understand why shorter packets have a better chance to "get through" while providing a higher amount of redundancy that further helps their recovery. In this project the focus is on five different sizes of text tags the string values are converted to the Hex values to be embedded into the songs; after generating the tags, the embedding process is started and the detail information is stored for the further analysis. Table 4.2 shows the tags' information.

4.2.3 Encoding Phase

There are two processes involved in this analysis which are embedding and extraction processes. Watermark is embedded into the selected audio files and the results for embedding process are depicted in Tables 4.3 for Rock, 4.4 for Pop, 4.5 for Classical, and 4.6 for Ambient. The arbitrary binary payload provided in tag is revolved into a data packet that comprises some synchronization, error checking, and the optional 12-bit time stamp information. Because of this

Table 4.4 Information of Embedding Process for Pop Songs				
	Tag A/Number of Packets	Tag B/Number of Packets	Tag C/Number of Packets	Tag D/Number of Packets	Tag E/Number of Packets
Song1	146	88	64	50	42
Song2	146	88	64	50	42
Song3	146	90	66	52	42
Song4	146	88	64	50	42
Song5	146	88	64	50	42
Song6	146	88	64	50	42
Song7	146	88	64	50	42
Song8	146	88	64	50	42
Song9	146	88	64	50	42
Song10	146	90	64	50	42

Table 4.5 Information of Embedding Process for Classical Songs				
	Tag				
	Tag A/Number of Packets	Tag B/Number of Packets	Tag C/Number of Packets	Tag D/Number of Packets	Tag E/Number of Packets
Song1	146	88	64	50	42
Song2	146	88	64	50	42
Song3	146	88	64	50	42
Song4	146	90	64	50	42
Song5	146	88	64	50	42
Song6	146	90	64	50	42
Song7	146	90	64	50	42
Song8	146	88	64	50	42
Song9	146	88	64	50	42
Song10	146	88	64	50	42

fixed overhead, the effective data rate at which marked your audio is a function of the payload size.

Tables 4.3–4.6 show that the number of packets for 5-byte tags is almost the same for all 40 songs and as it is seen the other four size tags are repeated extremely similar in all the songs and interestingly the packets are equally generated in four different genres of songs that it proves the difference between characteristics of various genre has not a role on embedding process. Another fact that can be realized from the table is that the higher the data payload the lower the number of packets.

Table 4.6 Information of Embedding Process for Ambient Songs					
	Tag A/Number of Packets	Tag B/Number of Packets	Tag C/Number of Packets	Tag D/Number of Packets	Tag E/Number of Packets
Song1	146	90	66	50	42
Song2	146	90	66	50	42
Song3	144	88	66	52	42
Song4	146	90	66	50	42
Song5	148	90	64	50	42
Song6	146	90	64	50	42
Song7	146	90	64	50	42
Song8	146	90	64	50	42
Song9	146	88	66	50	42
Song10	150	92	66	50	42

Table 4.7 Information of Extraction Process of Rock Songs					
	Tag A/Number of Packets	Tag B/Number of Packets	Tag C/Number of Packets	Tag D/Number of Packets	Tag E/Number of Packets
Song1	59	5	2	1	0
Song2	72	58	6	27	2
Song3	62	10	8	12	0
Song4	122	57	11	30	2
Song5	0	0	0	0	0
Song6	97	46	3	25	0
Song7	34	32	42	30	29
Song8	39	7	4	0	1
Song9	59	24	5	2	0
Song10	6	6	0	0	0

4.2.4 Decoding Phase

Successful extraction of a watermarked audio is the primary rule of each audio watermarking algorithm, because if the method cannot gain the original watermark from the audio successfully the usage of watermarking is meaningless. In this step of analysis all the water-marked songs will go through the same method of extraction to reveal whether the watermark can be extracted or not. In addition, there is no attack lunched to them yet. Tables 4.7–4.10 represent the result of extraction and show the number of packets that has helped to retrieve the information from audio files correctly. In the tables a gray color is used to highlight every cell that represent 0 for the number of packets,

Table 4.8 Information of Extraction Process of Pop Songs

	Tag A/Number of Packets	Tag B/Number of Packets	Tag C/Number of Packets	Tag D/Number of Packets	Tag E/Number of Packets
Song1	23	5	4	0	1
Song2	79	41	38	34	25
Song3	106	75	37	23	6
Song4	132	73	58	37	21
Song5	45	38	18	22	8
Song6	48	25	8	1	8
Song7	50	23	15	14	5
Song8	59	9	6	1	2
Song9	132	79	52	42	27
Song10	75	41	9	10	2

Table 4.9 Information of Extraction Process of Classical Songs

	Tag A/Number of Packets	Tag B/Number of Packets	Tag C/Number of Packets	Tag D/Number of Packets	Tag E/Number of Packets
Song1	82	55	14	19	4
Song2	87	36	10	13	3
Song3	84	52	17	23	11
Song4	83	29	0	20	1
Song5	117	44	31	35	7
Song6	108	59	29	29	11
Song7	137	76	44	39	27
Song8	64	26	2	12	1
Song9	91	68	40	33	23
Song10	58	27	6	13	3

which mean the watermark is not extracted at all there. By the end of this section there is one chart to represent the percentage of success for each specific group of songs specified in Fig. 4.1.

The table shows that Rock has difficulty to retrieve 25-byte tag successfully, and also with having the same audio size and embedding a bigger watermark the chance to extraction decreases directly.

From Tables 4.8 and 4.9, it is clear that Classical and Pop genres of music have less challenges to retrieve the watermark when compared with Rock group and it is strongly depending on the instrumental attributes of them. Another interesting result is the size of massage, which

Table 4.10 Information of Extraction Process of Ambient Songs					
	Tag A/Number of Packets	Tag B/Number of Packets	Tag C/Number of Packets	Tag D/Number of Packets	Tag E/Number of Packets
Song1	47	11	3	6	0
Song2	106	80	43	42	18
Song3	44	33	0	2	0
Song4	101	71	39	46	24
Song5	125	78	45	44	24
Song6	97	52	27	29	15
Song7	0	0	0	0	0
Song8	34	17	13	12	8
Song9	31	22	4	8	0
Song10	27	0	4	3	0

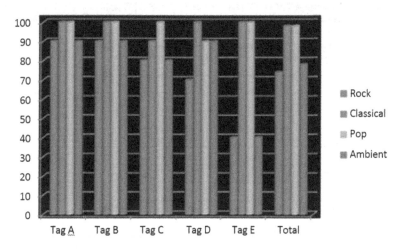

Figure 4.1 Proportion of successful extraction.

shows that it is impossible to retrieve a smaller massage, where the bigger one except these tow genres uses the method of watermarking that can generate a series of pseudorandom numbers to find the threshold of each frequency to select the sufficient one for distributing the watermark packets on them.

The results for Ambient songs is slightly similar to the Rock as it has challenges to extract the biggest tags more around 60%, after that the problem was for 15-byte size and this part proves the previous finding which was about the size of tags because it has two fails on 15-byte tag but one fail on 20-byte.

As Fig. 4.1 shows, the highest proportion of successful extraction is allocated to Pop and Classical equally after that Ambient and the last group which has more challenge to retrieve the packets is Rock.

4.3 ATTACKS

In this part after observing the proficiency of the selected method of audio watermarking, different types of attacks would be launched to the audio files to understand which type of attack is more threatening to the audio files. From Section 4.2, the results showed that there are two songs in this process which did not retrieve any tags at all, Song5 in Rock and Song7 in Ambient group. In next step the mp3 attack is launched to all 40 songs. It is obvious that the watermark would not be extracted after attack from those particular songs.

4.3.1 MP3 Attack

To know the characteristics of this type of attack that is very common refer to Chapter 2, Literature Review. The aim to practice this attack is to see how much resistance a watermarking method have on this very common conversion, mp3 format is the most popular one through the youth; however the quality is not as much high as in wav files but limit size and good quality of this type of audio files leads them to be the most shared type especially through the Internet. There are varieties of software that convert audio files to a different type and through the process of conversion some of the main parts of the song would be removed or changed totally and it can cause a really bad effect on a watermarked audio but because there are lots of different technical issues to produce a genre the effect of this conversion has not to be the same for all four genres. The final result can show the impact of mp3 conversion on each type of genera of music specifically.

To run this attack, first the watermarked audio wav files are converted to mp3 using "Aiseesoft MP3 WAV Converter" with audio bitrate 128 bkbps and audio sample 44,100 Hz. After that all the converted audio files are reversely renewed to wav files again similarly with audio bitrate 128 bkbps and audio sample 44,100 Hz.

Table 4.11 Unsuccessful Extraction After mp3 Attack					
	Tag A	Tag B	Tag C	Tag D	Tag E
Rock	3	1	2	6	9
Pop	0	0	0	0	2
Classical	0	0	2	8	0
Ambient	3	3	6	5	1
*Numbers are out of 10.					

Table 4.11 shows the number of unsuccessful extraction after conducting mp3 attacks in each category of songs and the following features are depicted to illustrate number of packet using to extract the watermark specifically.

Table 4.11 shows that Rock has near 50% failure after mp3 attack 21 out of 50 songs fails to retrieve the watermark and to compare this proportion with the previous one it is seen that after mp3 attack the amount of failure increased from 30% to 42%. After that Ambient group has failed 38%, 18 out of 40 songs and it is cleared that the amount of failure increased here too from 21% to near 40%, it almost got doubled, and the third place is for Classical with exactly 20% fails 10 out of 50 songs although from the extraction part it had only 3% failure. Pop songs were totally successful except in 25-byte tags with 2 out of 10 failures. It seems that this type of music can stand alone with mp3 attack better as the result is approximately the same as previous stage. In extraction it has failed only once and after mp3 attack just one more 25-byte watermark could not be retrieved. The results of successful extraction regarding to the number of packets that has been used to retrieve the watermark are shown in Fig. 4.2.

4.3.2 Resampling

In this part two types of resampling attack are launched to the selected songs. Resampling is the digital process of changing the sample or dimensions of digital imagery or audio by temporally or a really analyzing and sampling the original data. Audio resampling is also called sample rate conversion. This operation in digital signal processing involves converting a sampled signal from one sampling frequency to another. For instance, the output waveform of a digital audio workstation that was processed at 96 kHz must be resampled to 44.1 kHz to be placed on a Compact Disc. In commercial use of sound a lot of

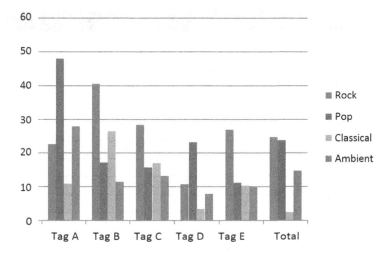

Figure 4.2 Successful extraction using number of packets.

signal processing is done to make the audio files suitable for specific application and during these changes some of the frequencies would be changed and some packets of watermarks would be lost or compromised. Resampling operation in digital signal processing involves converting each sample of a sampled signal from one depiction of amplitude to another. For occurrence, the output waveform of a digital audio workstation that was processed using 24 bits to represent amplitude would be resampled to 16 bits to be placed on a Compact Disc. This operation in digital signal processing involves changing the sampling rate of a discrete-time signal to obtain a new discrete-time representation of the underlying continuous-time signal. In Sections 4.3.2.1 and 4.3.2.2 two common types of resampling (down sampling and upsampling) is done for one song in each genre of music.

4.3.2.1 Down Sampling

In signal processing, down sampling (or "subsampling") is the process of reducing the sampling rate of a signal. In this stage of project audacity is used to do down sampling, the default sampling rate is 48,000 Hz, through the analysis it will decrease to 32,000, 22,050, 16,000, and 11,025 Hz each genre sample can resist to how much degradation of sampling rate.

Table 4.12 depicts that all the songs are resistant to the 32,000 and 22,050 Hz but for 1600 Hz Rock and Classical have problem with tag size 15 byte and above however the Pop song only have problem with

Table 4.12 Extraction After Down Sampling Attack					
	Tag A	Tag B	Tag C	Tag D	Tag E
Rock					
32,000 Hz	124	57	15	30	2
22,050 Hz	22	69	29	32	12
16,000 Hz	16	3	0	0	0
11,025 Hz	0	0	0	0	0
Classical					
32,000 Hz	80	55	15	17	4
22,050 Hz	87	61	12	17	3
16,000 Hz	14	4	0	0	0
11,025 Hz	0	0	0	0	0
Pop					
32,000 Hz	27	8	32	27	21
22,050 Hz	129	82	58	42	31
16,000 Hz	102	40	0	13	0
11,025 Hz	0	0	0	0	0
Ambient					
32,000 Hz	101	71	40	40	18
22,050 Hz	99	80	42	46	17
16,000 Hz	55	33	6	15	2
11,025 Hz	11	6	1	2	1

15-byte. From this result, it is understand that small size tags are not always more sufficient when compared with bigger size tags and it is completely dependent on what type of procedure does the application using to distribute packets through the song and how it calculates the possible threshold to insert the packet on their picks. The most significant part of the result is for ambient song. It is totally successful for all the decreases and even by only one packet it can retrieve the water mark after down sampling (Fig. 4.3).

4.3.2.2 Upsampling

In this step the reverse process of down sampling would be run on audio files named upsampling which is the procedure of growing the sample rate of a digital audio signal. For illustration, the sample rate of a CD is 44.1 kHz, which means the analog signal is sampled 44,100 times per second. Upsampling by a factor of two increases the sample rate from 44.1 to 88.2 kHz, effectively doubling the number of samples

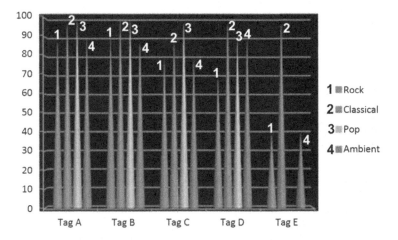

Figure 4.3 Proportion of successful extraction after down sampling.

available. The benefit of upsampling the signal widens frequency response. The maximum frequency response of a signal sampled at 44.1 kHz is 22.05 kHz per channel, while a signal sampled at 88.2 kHz has a maximum frequency response of 44.1 kHz per channel.

It is interesting that after upsampling all the song from 48,000 to 88,200 Hz and 1,920,000 Hz and also a very high one 3,840,000 Hz, there was no problem to extract all the embedded watermarks from all the songs, it showed that this king of attack is not harmful for all these genes.

4.3.3 Low-pass filtering

Filtering is so important both in sound and radio implementations. And all of this is just to get rid of inessential noise or signals that happen to be bombarding the inputs. So let us take a more in-depth look at these. Low-pass filters do what they say, allow low frequencies to pass by. These are sometimes known as treble cut filters because they lower the amplitude/oomph of higher frequencies. So if you look at Fig. 4.4, you will notice that as the frequency increases, the voltage/ amount of power decreases. To ensure that the sampling theorem is satisfied, or approximately so, a low-pass filter is used as an antialiasing filter to reduce the bandwidth of the signal before the signal is down sampled; the overall process (low-pass filter, then down sample) is sometimes called decimation.

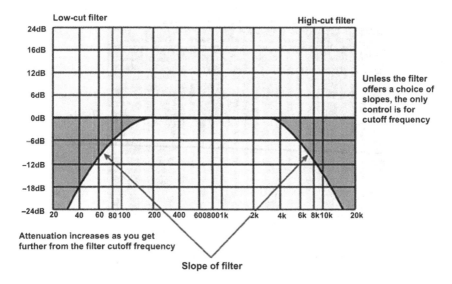

Figure 4.4 Representations of low- and high-pass filters (http://www.soundonsound.com).

Table 4.13 Extraction After Low-Pass Filter Attack					
	Tag A	Tag B	Tag C	Tag D	Tag E
Rock	28	0	3	3	2
Pop	82	60	50	22	23
Classical	9	7	4	0	0
Ambient	58	30	14	14	14

Table 4.13 illustrates the result of extraction after passing the songs from a low-pass filter with 6 db per octave and they had been catted off frequency 821 Hz. In this type of attack again Ambient songs are successful totally also Pop songs have no problem to extract the watermarks at all but in Classical group for tag size 20 byte and 25 bytes a failure has happened and in Rock song just 10-byte tag could not retrieve effectively. Classical songs already have a low frequency range so it is logical after cutting them down, the packets would be lost and became very hard or impossible to retrieve the watermark.

4.3.4 High-Pass Filtering
A high-pass filter is used in an audio system to allow high frequencies to get through while filtering or cutting low frequencies. A high-pass

Table 4.14 Extraction After High-Pass Filter Attack					
	Tag A	Tag B	Tag C	Tag D	Tag E
Rock					
6 db–1141 Hz	127	75	25	31	7
12 db–2121 Hz	128	60	15	33	0
Classical					
6 db–1141 Hz	89	51	16	19	2
12 db–2121 Hz	91	62	14	21	3
Pop					
6 db–1141 Hz	131	61	52	37	17
12 db–2121 Hz	130	69	43	38	13
Ambient					
6 db–1141 Hz	118	83	47	46	37
12 db–2121 Hz	120	73	39	46	9

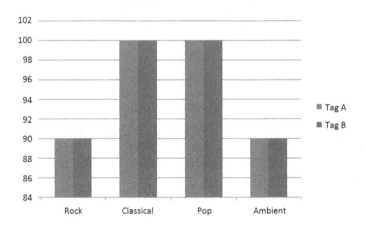

Figure 4.5 Proportion of successful extraction after filtering attack.

filter is used with small speakers to remove bass or low frequencies. In this step first a high filter with 6 db and 1141 Hz provided for all songs but as there was not any problem to extract this time experiments a higher power 12 db and 1221 Hz of frequencies. Table 4.14 shows the result of this filtering on the watermarked songs. It is significant that except one failure in 25-byte size tag from Rock group all the other songs successfully pass the extraction process. Also the comparison between the results of high-pass filter and low-pass filter is depicted in Fig. 4.5.

4.3.5 Echo

An echo is simply a slightly delayed repetition of a sound. In our acoustic environment, echoes are heard as the result of sound waves being reflected from a surface and returned to the listener. The time difference between when the listener hears the direct sound and when the reflection (echo) is heard is referred to as the echo delay time. This delay time depends on the distances between the listener, the sound source, and the reflecting surface, with greater distances resulting in longer delay times. In this project all the watermarked sounds were tested with delay time 1 second and decay factor 0.5. Table 4.15 spectacles the result of this process. As it is seen, Rock songs have the most challenges with this type of signal processing and it leads to fail to extract even 1 tag at all. Remarkably despite previous processing Pop song failed to retrieve all the tags except only 5-byte size using 2 packets. Ambient song had also problem to regain the 15-byte and 20-byte and 25-byte tags. The only complete effective group to face with this attack is Classical.

4.4 FINAL RESULT

The process of attack analysis with specific dataset gathered in this study is finished. Fig. 4.6 demonstrates an integrated result from all the attacks on all the genres and it shows which type of attacks is more treacherous and also which group of songs is more vulnerable than the others.

Table 4.15 Extraction After Echo Attack					
	Tag A	Tag B	Tag C	Tag D	Tag E
Rock	0	0	0	0	0
Pop	2	0	0	0	0
Classical	82	55	14	19	2
Ambient	10	2	0	0	0

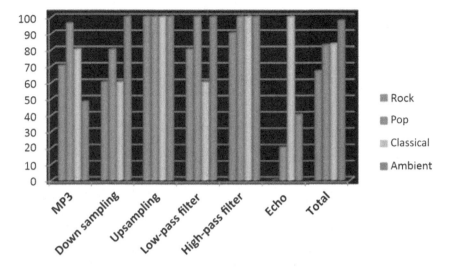

Figure 4.6 Proportion of successful extraction for all attacks.

4.5 SUMMARY

This chapter analyzed results of five different types of signal processing attacks on a specific watermarking algorithm. Results included five different sizes of text tags that inserted into variety of songs which belonged to four different groups of music. Through this analysis, the persistent of watermark in each group of songs has been tested to see which type of attack is more challenging for audios and to understand every genre of music is vulnerable to what kind of attack the most. The results showed that the genres are so important on the resistance and each one is so different from the other and has its own vulnerabilities. In addition, the most harmful attack is echo and the least is upsampling.

CHAPTER 5

Evaluation Analysis

5.1 INTRODUCTION

In this section the discussion about audio file analysis and evaluation is discussed, besides the most important types of comparison for two audios (host audio and the watermarked audio) represented. Also two different comparisons are done to precise what kind of attribute changes of audio has the most influential impact on the power ratio of the result sound of subtract the watermarked audio from host, to estimate SNR of it. Finally, there is a conclusion that briefly explains the entire results. Chapter 4, Attack Analysis, discussed about different types of attacks on audio files and the result showed that there are a lot of differences between the Pop, Rock, Classical, and Rock music.

Embedding some information to provide authentication of the owner of the content, e.g., the singer of a song or the related data about some images that can be visible or invisible, called watermarking is an old way to hold intellectual copyright protection and some other aims like "owner identification, proof of ownership, transaction tracking, and copy control" (Cox and Miller, 2002). But because there are so many threats in the network, a watermark is vulnerable to the alternation or even being removed from the host (the content that the watermark is embedded there) by the attackers, so the necessity of having new methods to make the watermark more robust is obvious. In addition, regarding to the audio type content there are numerous attacks that can alter the file and diminish the quality of songs by the way of different types of signal processing. Knowing all types of possible attacks on audio files and analyzing how they can effect on a digital audio content is a vital issue that can help the scientist to propose a well method of audio watermarking by considering all the characteristics of a song and the threats related to it.

Audio Content Security. DOI: http://dx.doi.org/10.1016/B978-0-12-811383-7.00005-2

5.2 CORRELATION BETWEEN POWER AND SIZE RATIO IN AUDIO WATERMARKING

In this section the experiments of embedding the different message into the same host are provided for two hosts with two different sizes. As Table 5.1 and Fig. 5.1 show, 12 messages with different sizes are embedded into four hosts whose size is 11 mb. The results of these experiments are compared with the previous study which has been done a partially similar analysis but not in audio watermarking, in audio steganography. The achieved results of this step can prove or

Table 5.1 Correlation Between PSNR and Size Ratio				
	Wav Size (MB)	Tag Size (Byte)	PSNR (Left) (db)	PSNR (Right) (db)
Rock	11	2	81.1	71.6
Rock	11	4	80.9	70.6
Rock	11	8	74.8	67.6
Pop	11	2	87.2	86
Pop	11	4	85.8	80.1
Pop	11	8	84	81.1
Ambient	11	2	109.4	109.3
Ambient	11	4	98.4	98.8
Ambient	11	8	97.5	976
Classical	11	2	83.6	79.7
Classical	11	4	82.1	80.4
Classical	11	8	78.8	76.8

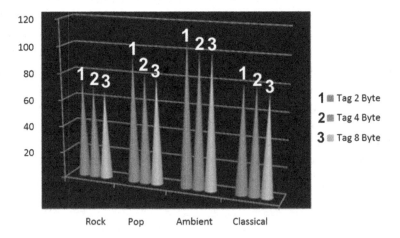

Figure 5.1 Representation of correlation between size and PSNR.

disapprove the last work that had been done. Also the study is on stereo mode means that audio output is on two channels and the peak signal-to-noise ratio (PSNR) are calculated for each separately.

The results of this stage showed that whenever the sizes of tags are increased the PSNR decreased. The next table and figure are used to test another audio with bigger size than previous one to preserve whether the results are the same as this one or not. As Table 5.2 and Fig. 5.2 show, 12 messages with different sizes are embedded into 4 hosts whose size is 16 mb.

Table 5.2 Correlation Between PSNR and Size Ratio 2				
	Wav Size (MB)	Tag Size (Byte)	Power (Left) (db)	Power (Right) (db)
Rock	16	2	74.1	7.9
Rock	16	4	72.8	66.3
Rock	16	8	69.1	63.2
Pop	16	2	103	80.7
Pop	16	4	84.8	80.7
Pop	16	8	81.9	78.7
Ambient	16	2	108.8	107.9
Ambient	16	4	100.5	99.7
Ambient	16	8	100.2	99.4
Classical	16	2	83	99
Classical	16	4	79.6	79.6
Classical	16	8	76.1	72.7

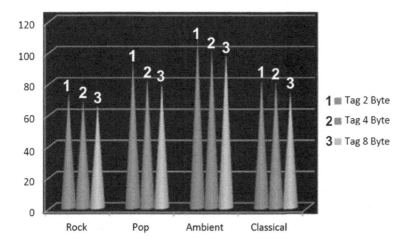

Figure 5.2 Representation of correlation between size ratio 2 and PSNR.

The results brought credits to Table 5.1 because Fig. 5.2 marches the same relation between size ratio and PSNR.

5.2.1 Previous Work

The experiment of this study depicted that there is a reverse relation between size of the massage that is embedded to the host and PSNR of them. Fig. 5.3 shows the results of previous study that gain the same results too but not with this method. Fifteen messages with different sizes are embedded into a host whose size is 126,215,470 bytes. The results showed that as the message size is increased, PSNR is decreased (Zamani et al., 2012). Besides this achievement the study aimed to answer this question: is it important what genre of music is tested or not? It means does the relation between size ratio and PSNR the same for all genre of music or not. The experiment showed that the relation is similar for all four genres generally; however, it differs regarding to the quantity of PSNR itself.

5.3 CORRELATION BETWEEN PSNR AND SAMPLING RATE IN AUDIO WATERMARKING

As Table 5.3 shows, four different genres of music files with wav format were selected. Then process was run with two different sampling

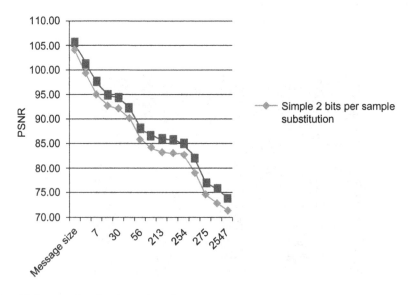

Figure 5.3 Representation of previous study result.

	Wav Size (MB)	Tag Size (Byte)	PSNR (Left) (db)	PNSR (Right) (db)	New PSNR (Left) (db)	New PNSR (Right) (db)
Rock	16	2	74.1	71.9	72.8	67.5
Pop	16	2	83	77.7	82.1	68.5
Ambient	16	2	81.8	78.9	79.4	76.4
Classical	16	2	83	79	72.8	67.5

Table 5.3 Correlation Between PSNR and Sampling Rate

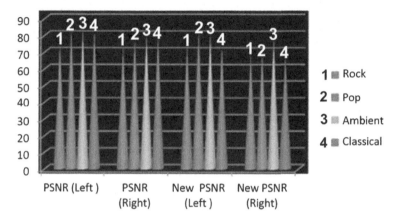

Figure 5.4 Representation of correlation between PSNR and sampling rate.

rates. This study shows that for tow-specified sampling rate almost tow-specified PSNR yields. In other words, if the size of message file is almost the same, the PSNR of embedding depends on the sampling rate of audio files. The reduction of sampling rate has a direct effect on PSNR means it is degraded too. Besides it shows that no matter what genre of music is the host, the result of change on sampling rate has totally the same influence on all (Fig. 5.4).

5.4 SUMMARY

The study's focus in this chapter is on the effect of different parameters of sound and tags on the PSNR of the host and watermarked audio. The idea is gotten from a similar study on audio steganography but the elements of tests are different. The test is conducted on stereo mode with left and right channel. First test is related to the size ratio between host and tag and the relation between PSNR and it. Throughout the analysis the results showed that there is a reverse relation between them which proves the previous study, but there is no difference between various genres of music and the correlation is almost the same for all; however, the number of PSNR is not the same. The second test is based on the sampling rate and its effect on PSNR and the results indicated that whenever the sampling rate decreased the PSNR decreased too. Finally, in this chapter the results showed that regarding to PSNR there is no matter what genres of music is the host the effective parameters are the same for all.

CHAPTER 6

Conclusion

6.1 INTRODUCTION

In this section the discussion about findings of the impact of attacks on Dual method in audio watermarking is discussed in further details. Based on the analysis Chapters 4, Attack Analysis, and 5, Evaluation Analysis, the findings would be explained here and an overlooked consideration is clarified; also the possible future works based on this project would be introduced too. Following sections discussed about how the study achieved to its objectives and meets its goals which were: (1) to analyze the possible signal processing on audio files and observe their effects on the embedded watermark and the audio itself regarding to different genres of music, (2) to correlate different parameters of audio structure and watermark on PSNR, and (3) to suggest a new consideration before producing a new method for audio watermarking.

6.2 FINAL ACHIEVEMENTS

In this study the main focus is to find the most important reasons that cause audio watermarking being overlooked in the academic and commercial world in comparison with other contents like image, text, and videos. One of the most important reasons is that audio signals are so vulnerable to different types of attacks and also because it deals with human auditory system and this system is so sensitive to the noise. Any kind of changes in the original signal could be heard by ears. During the literature review, the author found that although there are a lot of types of audio files and music styles, there is no specific method of watermarking for them. Also in commercial world there are a few people who are aware of the possibility of copyright protection using watermarking for audio; instead most of them believed that watermarking is used for images only. The analysis of Chapter 4, Attack Analysis, was based on this reason and tried to find out if there is any relationship between different genres of music and their

Audio Content Security. DOI: http://dx.doi.org/10.1016/B978-0-12-811383-7.00006-4

resistance in face of various kinds of attacks or not. The final result proves that depending on what sort of music is tested the effects of attacks are so different. However, the method of embedding and extraction for all of them was the same. The fact that one method can be secure against some attacks was discussed before but there is no discussion about security of audio watermarking for different genres of music. In Chapter 5, Evaluation Analysis, the analysis was based on the effects of size ratio between host and tag and also correlation between PSNR and sampling rate of the song regarding to different genres of musics. The results showed that the effect is the same for all four genres.

6.3 CONTRIBUTION OF STUDY

Following the achievements of this project, the author wants to suggest that consideration of sound genre characteristics would be so benefited before producing a novel method for watermarking. Also there is no watermarking method which is just belong to a particular genre of song, and regarding the analysis of this study, it would be very useful to have different methods of audio watermarking for different groups of songs. And it is really needed to protect the intellectual property for audio contents these days.

6.4 FUTURE WORK

In the future there is an open area to follow the structure of this study for different datasets and genres of music and also to run the project on other method of audio watermarking to see how many results have similarity to each other but with the same dataset of this project. Also future students can select another method of evaluation and test audio structure specification effects on that particular method.

REFERENCES

Atoum, M.S., et al., 2011. A steganography method based on hiding secret data in MPEG/audio layerIII. IJCSNS 11 (5), 184.

Chen, B., Wornell, G.W., 2001. Quantization index modulation methods for digital watermarking and information embedding of multimedia. J VLSI Signal Process Syst Signal Image Video Technol 27, 7–33.

Chen, S.-T., Huang, H.-N., Chen, C.-J., Tseng, K.-K., Tu, S.-Y., 2013. Adaptive audio watermarking via the optimization point of view on the wavelet-based entropy. Digit Signal Process 23 (3), 971–980.

Cox, I.J., Miller, M.L., 2002. The first 50 years of electronic watermarking. EURASIP J Appl Signal Process 2002, 126–132.

Deschamp, J., Guerrero, D., Zwiebel, R., 2012. Multi-Channel Audio Display. Google Patents.

Fallahpour, M., Megías Jiménez, D., 2012. High capacity robust audio watermarking scheme based on FFT and linear regression. Int J Innov Comput Inform Control 8 (4), 2477–2489.

Fallahpour, M., Megias, D., 2012. High capacity logarithmic audio watermarking based on the human auditory system. In: IEEE International Symposium on Multimedia (ISM) 2012. IEEE, pp. 28–31.

Gupta, S., Cho, S., Kuo, C.-C., 2012. Current developments and future trends in audio authentication. MultiMedia, IEEE 19, 50–59.

Hall, T., 2002. Music piracy and the audio home recording act. Duke Law Technol Rev 1, 1–8.

Jackson, W., 2013. An Introduction to Audio: Concepts and Optimization. Learn Android App Development. Springer, USA, pp. 321–344.

Kaur, H., Kaur, U., 2013. Blind audio watermarking schemes: a literature review. Eng Sci Technol Int J 3 (2), 288–295.

Lei, B., Yann Soon, I., Zhou, F., Li, Z., Lei, H., 2012. A robust audio watermarking scheme based on lifting wavelet transform and singular value decomposition. Signal Process 92, 1985–2001.

Li, J.-S., Hsieh, C.-J., Hung, C.-F., 2010. A novel DRM framework for peer-to-peer music content delivery. J Syst Softw 83, 1689–1700.

Lu, C.-S., 2005. Multimedia Security: Steganography and Digital Watermarking Techniques for Protection of Intellectual Property. IGI Global, USA.

Patil, M.M.V., Chitode, J., 2012. Audio watermarking: a way to copyright protection. Int J Eng 1.

Peng, H., Wang, J., Zhang, Z., 2013. Audio watermarking scheme robust against desynchronization attacks based on kernel clustering. Multimed Tools Appl 62, 681–699.

Steinebach, M., Dittmann, J., 2003. Watermarking-based digital audio data authentication. EURASIP J Appl Signal Process. 1001–1015, 2003.

Varodayan, D., Lin, Y.-C., Girod, B., 2008. Audio authentication based on distributed source coding. In: IEEE International Conference on Acoustics, Speech and Signal Processing, 2008. ICASSP 2008. IEEE, pp. 225–228.

Wang, M.-L., Lin, H.-X., Lee, M.-T., 2012. Robust audio watermarking based on MDCT coefficients. In: 2012 Sixth International Conference on Genetic and Evolutionary Computing (ICGEC). IEEE, pp. 372–375.

Yershov, A., Karpelcev, R., 2011. Unified evaluation system for audio steganography methods. Sci J Riga Tech Univ. Computer Sci 44 (1), 151–157.

You, Y.-l., Smith, W.P., Fejzo, Z., Smyth, S., 2013. Improving Sound Quality of Established Low Bit-Rate Audio Coding Systems Without Loss of Decoder Compatility. EP Patent 2,228,790.

Yu, G., Zuo, J., Cui, D., 2011. Performance Evaluation of Digital Audio Watermarking Algorithm Under Low Bits Rates. Web Information Systems and Mining. Springer, Berlin Heidelberg, pp. 336–343.

Zamani, M., Manaf, A.B.A., Abdullah, S.M., Chaeikar, S.S., 2012. Correlation between PSNR and bit per sample rate in audio steganography. In: 11th International Conference on Signal Processing (SIP'12), 2012, pp. 163–168.

Printed in the United States
By Bookmasters